E=mc2
Machu Pichu
KENNY'S
FAVORITE TRIVIA BOOK
CRAZY AND FUN WACKY FACTS!
BLACK & WHITE EDITION
Chick-fil-A
Kenny 'K3' Rochon, III & Dr. Smiley

Kenny's Favorite Trivia

Black & White Edition

By Kenny 'K3' Rochon, III

and Daddy

ISBN: 978-1-64810-101-4

PRINTED IN THE UNITED STATES OF AMERICA

This is dedicated my Mom and Dad and Grandpa and Grandma Ruthann Aunt Lizy, Uncle Alex, Caroline, Stephanie, Uncle Stevie, Aunt Sharron, Matt Rochon, Andrew Rochon, Aunt Jen, Uncle Patrick, Kay Kay Herwig, Brendan Herwig.

To my friends: Ally, Amelia, Anthony, Barbara, Beth, Cathy, Drake, Eddie, James, Jayden, Jonah, Nico, Tyson, Jackson, and all my friends at school.

And to my teacher Mrs. Jackson.

-Kenny

This is dedicated to my favorite people in the world. My son who lights up my life and inspires me to publish a book a year for him and my loving wife who is the best mother in the world to our son. Raising a child with love is the way to ultimately bring more smiles to the world.

Also, a special thanks to Carolyn Sheltraw for her help with layout. Al Granger for his hand in making another book (almost 200) a reality and a super fun book to help fund Kenny's education.

- Ken Rochon, Jr

AGRICULTURE

An **acre** of land is about the same as an American football field.

Agriculture is the single largest employer in the world.

About 12% of the land on earth is used for **agriculture**.

Bananas are the **number one** fruit crop in the world.

Bananas are **curved** because they grow toward the sun.

Bees contribute to more than $15 billion worth of crops every year through the act of pollination.

The **world's costliest coffee**, at $130 a point, is called kopi luwak. It is found in the droppings of a type of marsupial that eats only the very best coffee beans. Plantation workers track them and scoop up their precious poop.

The **forest industry plants** an average of 4 million tress each day: over **1.5 billion trees per year.**

The Mesopotamians build the first **simple irrigation system** around 7,000 B.C.

Farmers will have to grow 70% more food than what is currently produced to feed the **world's growing population** by 2050.

Seeds had to be scattered by hand until Jethro Tull's seed drill (developed in 1701) made it possible to plant seeds in a row.

AIRCRAFT

Aerophobia or Pteromerhanophobia are the fear of flying.

Black boxes aren't actually black. They are painted a bright orange, so they are visible amount the wreckage.

A Boeing 747 airliner is made up of **6,000,000 parts**.

A Boeing 747 is more fuel-efficient than your car.

Hartsfield-Jackson (ATL) in Atlanta, Georgia, is the **world's busiest airport** with more than 970,000 airplane movements a year. That also equals more than **260,000 passengers per day!**

The **longest flight** by distance is Singapore to Newark, New Jersey with a total of 9,534 miles.

Hijacking of aircraft wasn't illegal until 1961.

At any given hour, there are over **61,000 people airborne** over the United States.

The **shortest flight in the world** connects Westray and Papa Westray Islands in Scotland. The total flight is 1.7 miles long and lasts 57 seconds.

Statistically speaking, you have a 96% chance of surviving a plane crash.

A study conducted in 2013 by the BBC found that 56% of pilots fell asleep while flying and 29% had woken up to find that the copilot was also asleep.

AMERICA

The **White House**, in Washington, D.C., was initially gray, the color of its original stone. After the Way of 1812, during which it had been burned by Canadian troops, the outside walls were painted white to hide the smoke stains.

ANIMAL

More than **99.9 percent** of all the animal species that have ever lived on Earth were **extinct** before the coming of man.

Blue Whale fart bubbles are large enough to enclose a horse.

Camels have three eyelids to protect themselves from blowing sand.

Cats have better **memories** than dogs. Tests conducted by the University of Michigan concluded that while a dog's memory lasts no more than five minutes, a cat's can last as long as 16 hours – exceeding even that of monkeys and orangutans.

Elurophobia is the fear of cats.

There are an estimated 1,000,000 species of life living in the oceans. Only 1% have been cataloged.

Cheetahs cannon roar. They meow like house cats.

Chocolate can kill dogs. It directly affects their heart and nervous system.

Cynophobia is the fear of dogs

Elephants are the only animals that can't jump.

Hippopotamus milk is pink.

Humans are the most **dangerous animal** on Earth.

An adult **lion's roar** can be heard up to five miles away, and wans off intruders or reunites scattered members of the pride.

An **ostrich**'s eye is bigger than its brain.

Sharks kill fewer than 10 people a year. In contrast, humans kill over 100 million sharks a year.

The **world's smallest mammal** at 1.1 inches, is the Kitti's hog-nosed bat.

The **toughest creatures on Earth** at just a millimetre or less in size, are tardigrades. A recent study found that they could survive almost any cosmic disaster that could hit the planet. Tardigrades are often found in locations that dry out such as in moss and in ponds.

Snails take the **longest naps**, some lasting as long as three years.

Zoophobia is the fear of animals

ARCHAEOLOGY

The **Pyramids of Giza** are the oldest and sole remaining intact wonder of the Seven Wonders of the Ancient World.

In the Quechua Indian Language, "**Machu Picchu**" means 'Old Peak" or "Old Mountain"

ARCHITECTURE

The **tallest building in the world** (2020) is the **Burj Khalifa** standing at 2,716.5 feet tall in Dubai. It is three times as tall as the Eiffel Tower and nearly twice as tall as the Empire State Building.

The base of the **Great Pyramids** in Egypt is large enough to cover ten football fields. According to the Greek historian Herodotus, it took 400,000 men twenty years to construct this great monument.

The Pagoda at Horyuji is the oldest wooden building on the planet. It contains some timbers that came from trees felled around 600 A.D.

ART

Most **Expensive** according to Guinness World Records lists Leonardo da Vinci's Mona Lisa as having the highest ever insurance value for a painting. On permanent display at the Louvre in Paris, the Mona Lisa was assessed at US$100 million on December 14, 1962.

The Louvre contains about 7,500 paintings, of which about 66% are by French artists. If you spent 30 seconds on each piece, it would take you 100 days to see every piece of art in the museum.

The Louvre is the most visited museum in the world.

Leonardo da Vinci's 'Mona Lisa' wasn't famous until it was stolen from the Louvre in 1911.

AVIATION

World's biggest plane is the wide body Antonov An-225 is powered by six turbofan engines and is the longest and heaviest airplane ever built, with a maximum takeoff weight of 640 tons. It was built to carry Soviet space shuttles 27 years ago.

Fastest jet aircraft in the world is the Lockheed SR-71 Blackbird reaching speeds of Mach 3.3 - that's more than 3,500 kph (2,100 mph) and almost four times as fast as the average cruising speed of a commercial airliner.

The Wright Brothers only ever flew together once, in 1910. This is because their father feared losing both sons in an airplane accident, so he gave them special permission for just that one time.

BODY

Human bone is ridiculously strong. A cubic inch of it can bear the weight of five standard pickup trucks making it four times as strong as concrete.

The human body has more than 650 muscles.

By the time you reach 50 years of age, you will have walked approximately 75.000 miles.

BOOKS

Bibliophobia is the fear of books.

World's **best-selling book of all time** is Migel de Cervantes' 'Don Quixote' selling more than 500 million copies. Second on the list published in 2017 is the Xinhua Zidian, the new Chinese Dictionary, with 400 million copies sold.

World's **largest book** according to the Guinness Book of World Records is given to the 2012 text on the Prophet Muhammad created in Dubai and measuring an impressive 16.40 ft x 26.44 ft.

BRAIN

Astoundingly the human brain, the universe, and the internet all have similar network patterns and the same growth dynamics.

A recent study found that a 45 - 60 minute nap during the day could improve learning and memory by as much as five fold. Researchers say that a **short nap** at the office or in school is enough to improve learning success.

It has been estimated that humans use only 10% of their brain.

BUSINESS

Apple originally had three co-founders. The third co-founder, a man named Ronald Wayne, had joined co-founders, Steve Jobs and Steve Wozniak but he left the company only 12 days after it was founded. He sold his 10% share for only $800 and got an additional payout of $1,500. Had Wayne stayed with the company, his shares would be worth today around $60 billion.

Apple is so successful, it has twice the amount of money than the U.S. Treasury.

The bite in Apple's logo is so it wouldn't be confused with a cherry.

First company to earn **$1 billion in one year** was General Motors, in 1955

Nabisco produces about **16 billion Oreo cookies a year** at its Chicago factory alone

The **busiest Pizza Hut** in the world is in Paris, France

Kellogg's makes more than **two billion Pop-Tarts** sold each year making it their most popular product.

Wal-Mart's annual income is nearly equal to that of Russia

Colonel Sanders had trouble getting his famous chicken just right. The secret recipe was rejected over 1,000 times before a restaurant finally added it to the menu.

Dasani Water is just purified tap water, it doesn't come from a natural spring.

Google and Alphabet (a subsidiary of Google) own over 200 companies.

CARS

Most **expensive** car ever sold is a 1962 Ferrari 250 GTO that sold at an RM Sotheby's auction for $48.4 million in 2018.

World's **fastest** Production Car is the 2020 SSC Tuatara Hypercar reaching a speed of 316.11 mph.

COLORS

Melanophobia is the fear of the color black

Porphyrophobia is the fear of the color purple

Leukophobia is the fear of the color white

Almost 95% of a car's lifetime is spent parked.

COUNTRIES

The only nation whose name begins with an "A", but doesn't end in an "A" is Afghanistan.

In Albania, nodding your head means no, and shaking your head means yes.

Emus and kangaroos cannot walk backward and are on the Australian coat of arms for that reason.

Instead of its Gross National Product, **Bhutan** measures its development and prosperity by its **Gross National Happiness**. It's been doing this since 1971 and is the only country to do so.

The most **dangerous** countries in the world are Libya, Syria and Afghanistan according to global medical and security specialists International SOS, which has assigned destinations one of five risk categories based on the threat posed by political violence (including insurgency, politically motivated unrest, terrorism and war).

Finland has been named the **happiest place in the world** for a fourth year running, in an annual UN-sponsored report. The **World Happiness Report** saw Denmark in second place, then Switzerland, Iceland and the Netherlands.

Iraq is the only country that ends with 'q'.

You can rent the entire country of **Liechtenstein** for 70,000 dollars a night. In fact, Snoop Dogg once tried to do that in 2011 for a video shoot but the country said no primarily because he just didn't give them enough notice.

Richest country in the world is Luxembourg with a GDP per capita of over $109,000.

There are more **bicycles** than people in Copenhagen, Denmark. And five times as many bicycles as cars. Over 50% of city dwellers commute by bicycle and over 41% arrive at their place of work or study by bike.

Panama is the only place in the world where one can see the sunrise in the Pacific Ocean and set on the Atlantic.

Every December 25th, the inhabitants of the Chumbivilicas Province in **Peru** celebrate Takanakuy. This is where men settle grudges from the past year by calling each other out and having a fistfight. Then everybody goes drinking to numb the pain and move on to a new year.

Qatar (pronounced 'Cutter') is the only country in the world to begin with the letter "Q".

Switzerland is the only country with a **square flag**.

India has the most vegetarians in the world.

CULTURAL

In Egypt around 1500 B.C. a shaved head was considered the ultimate in feminine beauty. Egyptian women removed every hair from their heads with special gold tweezers and polished their scalps to a high sheen with buffing cloths.

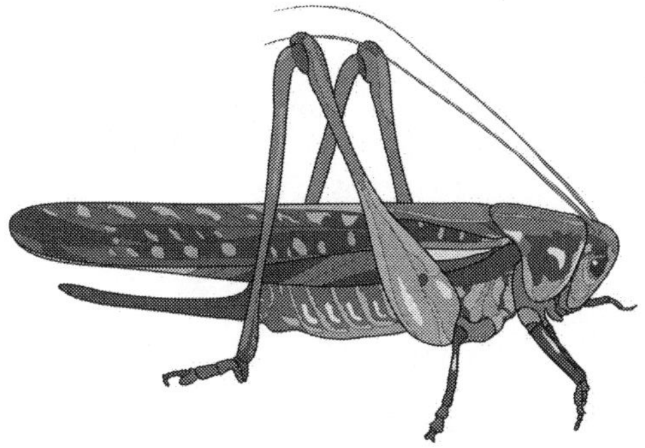

DANGEROUS

Most **dangerous dinosaur** to walk, or swim the Earth was the Spinosaurus.

Deadliest Insect are locusts While they do not directly kill humans, the devastation that an invasion can cause has been responsible for crop devastation and the consequential starvation of millions in history.

World's **Deadliest Spider** is the Brazilian Wandering Spider is a large brown spider similar to North American Wolf Spiders, but bigger and possessing a more toxic venom. It has the most neurologically active venom of all spiders and is regarded as the most dangerous spider in the world.

World's **deadliest** large land **mammal**, the hippopotamus killing an estimated 500 people per year in Africa. Hippos are aggressive creatures, and they have very sharp teeth. And you would not want to get stuck under one; at up to 4,000 pounds, they can crush a human to death.

World's most **painful sting** is the bullet ant. Although nonaggressive, bullet ants will sting when provoked. When one ant stings, it releases chemicals that signal other ants in the vicinity to sting repeatedly. The bullet anthas the most painful sting of any insect, according to the Schmidt Pain Index.

EARTH

Earth is moving through space at 66,487.72 mph.

The equatorial diameter of the Earth is 7,926 miles.

Light travels from the Sun to the Earth in about 8 minutes and 19 seconds.

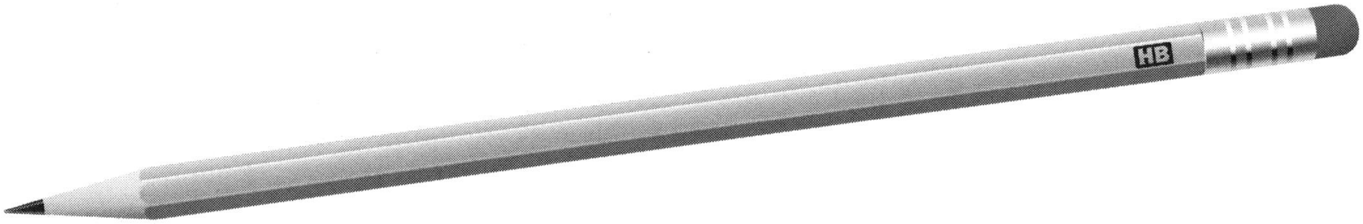

EDUCATION

Billionaire Elon Musk didn't like the school that his kids were attending, so he conceptualized and built his own named Ad Astra, which means 'to the stars'.

The average lead pencil will draw a line 35 miles long or write approximately 50,000 English words. More than 2 billion pencils are manufactured each year in the United States. If these were laid end to end they would circle the world nine times.

Scolionophobia is the fear of school

ENGINEERING

In 1452, a Hungarian engineer, known as Orban, offered to sell an extremely powerful canon to the Roman Emperor to the Roman Emperor. He refused, so Orban instead sold the cannon to the Ottoman Emperor, who used it to breach the walls of Constantinople in 1453, which brought the end of the Roman Empire.

ENTERTAINMENT

There's an annual conference called "The Boring Conference" on boring things. In the past, people have given talks on barcodes, sneezing and the sounds of vending machines.

Coulrophobia is the fear of clowns

The movie 'Straight Outta Compton' didn't play in Compton because there are no theaters there.

Watching your favorite movie over and over is good for you. The repetition calms you because knowing the outcome of a story helps you to feel safe in an unpredictable world. It also helps comfort you be recapturing lost feelings, according to a study in 2012 by Cristel Antonia Russell and Sidney J. Levy.

ENTREPRENEURS (ENTREPRENOLOGY)

The pet rock became so popular when it came out that within the first six months of its release, Gary Doll the inventor earned $15 million.

ENVIRONMENT

U.S. highway congestion costs over 160 billion a year, including wear and tear on vehicles, gas burned while idling, and lost productivity.

27,000 trees are cut down each day to make toilet paper.

Rainforests are cut down at a rate of 100 acres per day.

5,000 people die every day as a result of drinking unclean water.

EXPRESSIONS

The term "**dog days**" dates back to Roman times, when it was believed that Sirius, the Dog Star, added its heat to that of the sun from Sirius, the Dog Star, added its heat to that of the sun from July 3 to August 11, creating exceptionally high temperatures. The Romans called the period dies caniculares, or "days of the dog."

Clans of long ago that wanted to get rid of their unwanted people without killing them used to burn their houses down – hence the expression "**to get fired**".

The phrase "**sleep tight**" originated when mattresses were set upon ropes woven through the bed frame. To remedy sagging ropes, one would use a bed key to tighten the rope.

FASHION

In 4000 B.C., Egyptian men and women wore glitter eye shadow made from crushed shells of beetles.

FARTS

See Animals: Blue Whales ☺

In the Middle Ages, doctors believed that farting into jars and sniffing them, would prevent death or more specifically the black death plague which they called therapeutic stink.

Pumpaa was the first-ever character to fart in a Disney movie.

FAST FOOD

70% of the time ice from fast food restaurants was dirtier than toilet water

McDonald's is the largest toy distributor.

Chicken Nuggets contain very little meat and are mostly bird bones, fat and tendons.

A study found that 48% of soda fountains contain coliform bacteria, commonly found in feces.

Subway's bread used to contain a chemical also found in yoga mats, azodicaronamide.

Taco Bell was sued for calling their taco filling "meat", because it contains more oats, seasoning and filler than meat.

Those who eat fast food four times or more a week are at an 80% higher risk of dying from heart disease.

FITNESS

80% of Americans who have a gym membership don't use it.

Eugine Sandlow is considered the "Father of Bodybuilding"

In 1946, Joe Weider created the International Federation of Bodybuilders (IFBB) along with his brother Ben and started Mr. Olympia in 1965. The first contest was won by Larry Scott, who was awarded a bronze statue of Eugene Sandow.

FOOD

The **largest item** on any **menu** in the world is probably the roast camel, sometimes served at Bedouin wedding feats. The camel is stuffed with a sheep's carcass, which is stuffed with chickens, which are stuffed with fish, which are stuffed with eggs.

Grossest cheese is Casu Marzu made in Italy. Literally translated as "maggot cheese", Casu marzu is the world's grossest cheese. The Sardinian specialty is made from Pecorino cheese that intently becomes the natural breeding grounds for maggots.

Corn does not naturally exist in the wild. It is a human invention.

If humans consumed food like **cows**, they would have to eat 360 cheeseburgers and drink 400-800 glasses of water per day.

Cucumbers are 96% water.

The word Dorito means "little golden things" in Spanish.

It takes one hive of bees 55,000 miles of flight to produce one pound of **honey**.

Honey never spoils.

In the early 1900s, Mary Isabel Fraser visited China and brought back seeds to New Zealand. She grew the first crop of Kiwi in 1910. Today, New Zealand produces 1/3 of the world's supply of kiwi.

There are an average of 540 **peanuts** in a 12-ounce jar of peanut butter.

Arachibutyrophobia is the fear of peanut butter getting stuck to the roof of your mouth and choking.

World record for the **largest pizza** ever made clocks in at an amazing 26,833 lbs and was 122' 8" in diameter. This enormous pie was made in Norwood, South Africa on December 8, 1990.

Potatoes were the first vegetables ever planted in space.

Potatoes have almost all the nutrients that humans need to survive. To prove this, the executive director from the Washington State Potato Commission, ate nothing but potatoes for 60 days and was just fine.

In the USA, a pound of potato chips costs two hundred times more than a pound of potatoes.

Saffron, made from the dried stamens of cultivated crocus flowers, is the most expensive cooking spice.

Strawberry contains about 200 seeds on its surface and is the only fruit with seeds on the outside of the fruit.

The **heaviest strawberry** weighs 250 g (8.82 oz), which was grown by Koji Nakao (Japan) and was weighed in Fukuoka, Japan, on 28 January 2015. The strawberry was of a Japanese variety called Amaou.

Strawberries aren't really berries, because berries only have seeds on the inside, and a strawberry has its seeds on the outside.

Tomato is a fruit, not a vegetable

Canned white tuna has three times mores mercury than chunk light.

Cotton candy was invented in 1897 by a dentist. At the time, it was called 'Fairy Floss'.

A fresh egg will fall faster to the bottom of water. A bad egg will float and should be thrown out.

Fortune cookies were invented in San Francisco in the early 1900s.

Goat meat is the most popular meat worldwide. It accounts for 75% of the meat eaten globally.

Per ounce, nutritious food costs 10 times more than junk food.

Tonic water glows in the dark. Quinine is the component of tonic water that causes it to glow.

GAMES

Each King in a deck of playing cards represents a great king from history: Spades – King David, Clubs – Alexander the Great, Hearts – Charlemagne, Diamonds – Julius Caesar.

Minecraft's Endermen language is just English played backwards.

More Monopoly money is printed in a year, than real money throughout the world.

Candy Crush app brings in $633,000 a month in revenue

Contemporary playing cards were invented in France in 1480.

The player who goes first in Connect Four can win 100% of the time.

Poker is the most lucrative sport in the world. High roller tournaments awards total up to $44 million.

Risk was invented by Albert Lamoisse, a famous French film director best known for his short film "The Red Balloon".

It took Hungarian Inventor, Erno Rubik a month to solve his own invention.

GENDER (Men vs. Women)

Philematophobia is the fear of Kissing

Philophobia is the fear of love.

Gamophobia is the fear of marriage.

Gynophobia is the fear of women.

Venustraphobia is the fear of beautiful women.

GEOGRAPHY

Africam is a website that brings you live 24/7 video feed of protected areas of southern Africa. You can watch elephants, giraffes, hyenas, zebras, and all kinds of species of wildlife living there.

Istanbul, Turkey has a vending machine that releases food and water for stray dogs in exchange for recyclable bottles.

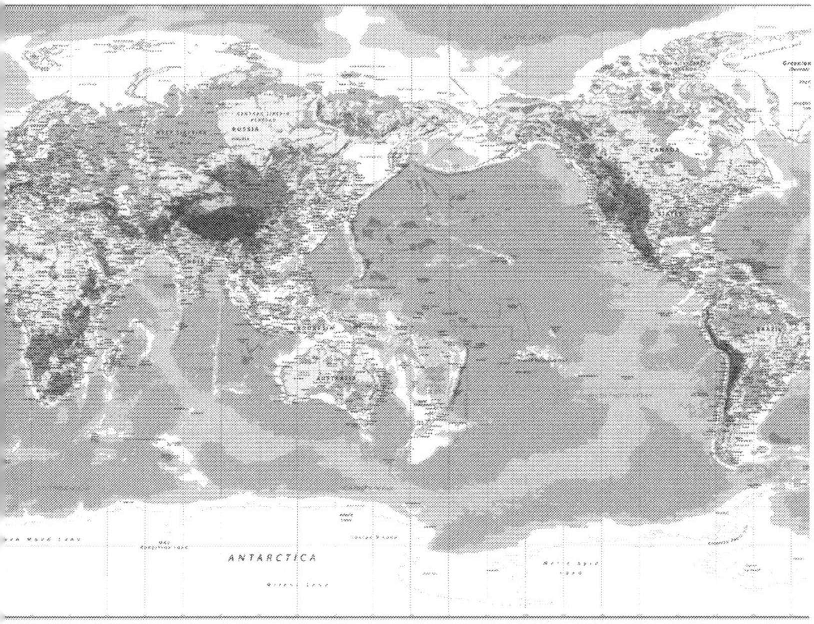

The first city to reach a population of one million people was Rome, in 133 B.C. London reached the mark in 1810, and New York made it in 1875. Today, there are over 300 cities in the world that boasts a population in excess of one million.

GOOGLE

Googling 'do a barrel roll' makes your whole browser flip over.

HAPPINESS

See Countries: Bhutan ☺

HEALTH

Exercising increases productivity. Having a regular exercise routine can make you happier, smarter and more energetic.

The first toothbrush with bristles was developed in China in 1498. Bristles were taken from hogs at first, later from horses. The nylon bristles were developed in 1938 by DuPont.

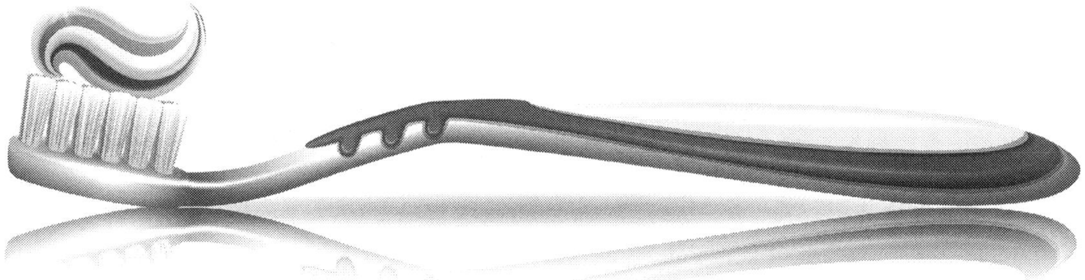

HYGIENE

If you chew on roasted coffee beans, it can help prevent the bacteria that causes bad breath.

HUMANS

The average **human body** contains enough iron to make a six-inch nail, sulfur to kill all the fleas on an average dog, carbon to take 900 pencils, potassium to fire a toy cannon, fat to make seven bars of soap, phosphorus to make 2,200 match heads, and water to fill a ten-gallon tank.

Babies are born without **kneecaps**. They don't appear until the child reaches 2 – 6 years of age.

Beards are the **fastest growing hairs** on the human body. If the average man never trimmed his beard, it would grow nearly 30-feet long in his lifetime.

World's **fastest** man is the Jamaican sprinter, Usain Bolt with an average ground speed was 37.58km/h, whilst reaching a top speed of 44.72km/h in the 60-80m stretch.

Srinivasa Gowda is a Kambala jockey from Mudbidri, Karnataka. He is also known as Indian Usain Bolt, Gowda came into limelight after his Kambala sprint broke Usain Bolt's world record in hundred meters. Gowda reportedly ran 142.5 metres in 13.62 seconds along with his racing buffalo pair.

The **heaviest human** ever recorded was Jon Minnoch and weighed 1,400 pounds (635 kg).

There is a study concluded by the journal of positive psychology that two-thirds of humans have **no idea** what they are good at and what their strengths are.

The **oldest** person living currently and the oldest person living (female) at the ripe age of 117 years and 41 days (verified on 12 February 2020) is Kane Tanaka (Japan, b. 2 January 1903).

World's **rubberiest** person is Daniel Browning Smith, also known as The Rubberboy (born May 8, 1979), is an American contortionist, actor, television host, comedian, sports entertainer, and stuntman, who holds the title of the most flexible person in history, owning a total of seven Guinness World Records.

World's **shortest** man, Khagendra Thapa Magar, who was born October 14, 1992, stood at 67.08 cm (2 ft 2.41 in) tall when measured at Fewa City Hospital in Pohkara, Nepal, on the advent of his 18th birthday in 2010. Guinness World Records is sad to hear of the passing today of the world's shortest man on January 17, 2020.

The **smartest person** in the world is writer Marilyn vos Savant (born 1946) has an IQ of 228, one of the highest ever recorded. Someone with a "normal" intelligence will score somewhere around 100 on an IQ test.

The **smartest man** in the world is Christopher Michael Langan (born March 25, 1952). He is an American horse rancher and autodidact who has been reported to score very highly on IQ tests. Langan's IQ was estimated on ABC's 20/20 to be between 195 and 210.

You can't usually **smell** your own house or perfume because of the survival instinct called "Olfactory Adaptation" the brain is always looking for new and unusual smells as a sign of possible danger, so it ignores the smells that it is familiar with.

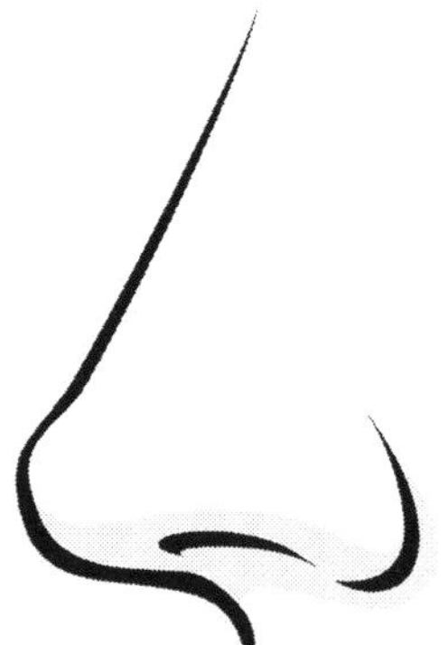

Gary Turner is a sideshow performer who holds the current Guinness World Record for the ***stretchiest*** *skin, caused by a serious form of Ehlers Danlos syndrome. On October 29, 1999, in Los Angeles, he stretched the skin on his abdomen to a total length of 6.25 inches, earning him the record.*

World's ***Strongest*** *Man is Ukraine's Oleksii Novikov after 1,185-pound partial deadlift.*

Žydrūnas Savickas is a Lithuanian powerlifter and professional strongman. Due to his multiple championship wins in all major international strongman competitions, as well as his longevity in the sport, he is widely considered by many to be the ***greatest strength athlete of all time****.*

The ***tallest*** *man "of whom there is irrefutable evidence" is Robert Wadlow. When last measured on 27 June 1940, the mild-mannered American stretched a staggering 2.72 m (8 ft 11.1 in) tall.*

You can tell identical ***twin babies*** *apart by their naval since the naval is a scar and isn't caused by genetics.*

A person will ***walk*** *the equivalent of 5 times around the equator during an average lifetime.*

INSECTS

Entomophobia is the fear of insects.

There are approximately 10 quintillion insects on the planet. That's 10 with 18 zeros behind it or roughly 200 million insects per human.

INVENTIONS

To have your picture taken by the very **first camera,** you would have to sit still for **8 hours!**

The **can opener** was invented 48 years after the can.

The **household wrench** was invented by boxing heavyweight champion Jack Johnson in 1922

The **parachute** was invented by Leonardo da Vinci in 1515.

See Entrepreneurs: Gary Doll: Pet Rock.

Potato Chips were invented by **George Crum** in **1853** near Saratoga Springs, New York. Crum was fed up with a customer complaining the chips were too soggy, so he cut a potato extra thin and fried them in hot grease and doused them with solt. The customer loved them, and the 'Saratoga Chips' were born.

The **quick-release ratchet** was designed by a teenager Peter M. Roberts. Sears told him the patent was valueless and gave him **$10,000,** then went on to make over **$44 million** selling the ratchets.

Scissors were invented by Leonardo Da Vinci

The **screwdriver** was invented before the screw.

Swim fins were invented by **Benjamin Franklin** when he was only **11 years old.** He was later recognized by his induction into the International Swimming Hall of Fame.

The **upside down ketchup bottle** earned its inventor **Paul Brown** **$13 million.**

Velcro was invented by a Swiss who was inspired by the way burrs attached to clothing while walking his dog.

White Out was invented by the mother of Mike Nesmith of the Monkees.

Windshield wiper was invented by Mary Anderson and patented on November 10th, 1903.

Robert Kearns invented the **intermittent windshield wiper** and saw his product being used on new cars. He sued Ford and received **$30 million.** Watch movie 'Flash of Genius'

LANDMARKS

*The **Great Wall of China** stretches over 5,500 miles.*

*There are 1,792 steps to the top of the **Eiffel Tower**.*

*In 1915, a millionaire named, Cecil Chubb bought his wife **Stonehenge** for approximately 10,000 dollars. Except that she didn't like it, so in 1918 he gave it to Britain and today it's valued at around 51 million dollars.*

***Stonehenge** is over 5,000 years old. Some of the giant stones were transported over 150 miles.*

MATH

***Arithmophobia** is the fear of numbers*

There is an app that exists called "Photo Math" that solves any question that you can point your phone at.

If one spells out numbers, they would have to count to One Thousand before coming across the letter "A".

MILITARY

Armored knights raised their visors to identify themselves when they rode past their king. This custom has become the modern military salute.

MONEY

In a real world transaction involving Bitcoin a man named Laslow, a computer programmer paid 10,000 Bitcoin for two large Papa John's pizzas. Only four years later those same Bitcoin were worth 5.12 million dollars and currently (2021) it is valued at $500 million.

Only 5% of Norway's financial transactions are done in cash. It is predicted that Norway will be ***cash free*** *very soon.*

World's Richest Person *is Amazon CEO Jeff Bezos with net worth of 191 Billion. Honorable mention is Elon Musk 166 Billion.*

The Lincoln Penny was first issued in 1909 to commemorate Abraham Lincoln's 100th birthday. The penny is the only coin where the figure faces to the right. This is because it was the direction he was facing in the photograph Victor David Brenner used to make his bas-relief of the penny.

Pennies in a garden will repel slugs which get electric shocks from touching copper and zinc.

There are 293 ways to make change for $1.

All 50 states are listed across the top of the Lincoln Memorial on the back of a $5 bill.

A farm in Delaware mulches more than four tons of worn-out U.S. cash into compost every day.

MUSIC / MUSICIANS

Blink-182 incorporated under the name Pennywise Poo Poo Butt, Inc. so that their accountants, managers, and attorneys would have to say that when doing business.

None of the Beatles could read music.

The British Navy uses Britney Spears songs to scare off the Somali pirates.

Coke La Rock is considered the first rapper in 1973.

ELO's first album "The Electric Light Orchestra" has a different name in the U.S. The American record company tried to call and confirm the name of the album. When they failed to reach anyone on the phone, the wrote 'No Answer', which someone misconstrued as the name of the album.

Elvis Presely is the best-selling solo artist of all time. Over one billion Elvis albums have been sold worldwide.

Bob Marley and Queen never won a Grammy while Milli Vanilli did.

Metallica is the first and only band to play on all seven continents.

One of the highest-attended concerts in history was held by Metallica on September 28, 1991 at Tushino Airfield in Moscow where 1.6 million people attended.

Michael Jackson bought the rights to most of The Beatles music for $47.5 million in 1985. it is now worth $450 million.

Michael Jackson cast 80 members of the Los Angelese Crips and Bloods for the music video "Beat It" in hopes to foster peace between them.

51.5% of deaths for hip-hop and rappers are homicide.

Scatman John suffered from a severe stutter from the time he learned to speak, which led to an emotionally traumatic childhood. He then made the song, "Scatman", in which he intended to inspire children who stuttered to overcome the condition. It reached #1 in 13 countries.

The U.S. economy loses $12.5 billion in total output annually as a consequence of music theft. Sound recording piracty leadds to the loss of 71,060 jobs in the U.S. economy.

Warner Music collected almost $2,000,000 a year for public usage of the song "Happy Birthday". However a judge ruled the song as "public domain" and they can no longer collect royalties on it.

The only guy without a beard in ZZ Top is Frank Beard.

NATURE

The Gippsland Lakes in Victoria, Australia glows in the dark. The light is created by a chemical reaction called bioluminescence when naturally occurring microorganisms in the water are disturbed.

A poisonous lagoon at the quarry in Buxton, England was dyed black for safety. The previous beautiful, azure blue watercolor was so inviting that the signs did not keep swimmers out.

Hydrophobia is the fear of water

NUMBERS

Octophobia is the fear of the figure 8

PEOPLE

United Kingdom engineer, **Eric Arthur Johnson** invented the **touchscreen** in the **1960s.** This would pave the way for the iPhone.

Charlie **Chaplin** once won third prize in a Charlie Chaplin look-alike contest.

Leonardo da Vinci could write with one hand and draw with the other hand at the same time. Despite his great scientific and artistic achievement, he was proudest of his ability to bend iron with his bare hands. He was also dyslexic. So strangely, not because of his dyslexia, he chose to write all of his notes to himself backward in mirror writing, meaning that you can only read them in the reflection of a mirror. However, notes that he wrote to other people, he wrote in normal direction. **He never built the inventions he designed!**

Leo Fender who invented the Telecaster and Stratocaster guitar, could not play guitar.

Benjamin Franklin never patented any of his inventions because he believed it was an opportunity to serve others freely and generously. He also invented the harmonica and it became so popular in Europe that Mozart and Beethoven composed music for it.

King Camp Gilette spent 8 years trying to invent and introduce his safety razor in 1901 and patented in 1904.

Arnold Shonberg suffered from Triskaidekaphobia, the **fear** of the number **13**. He died at 13 minutes from midnight on Friday the 13th.

Ruth Wakefield, the inventor of the **chocolate chip cookie** sold the idea to Nestle toll House in return for a lifetime supply of chocolate.

The **Wright Brothers** only ever flew together once, in 1910. This is because their father feared losing both sons in an airplane accident, so he gave them special permission just that one time.

PHOBIAS

Phobophobia is a fear experienced before actually experiencing the fear of the feared phobias its somatic sensations that precede it, which is preceded by generalized anxiety disorders and can generate panic attacks. Like all the phobias, the patients avoid the feared phobia in order to avoid the fear of it.

Acrophobia is the Fear of heights.

Chronophobia is the fear of Time

Koumpounophobia is the fear of buttons.

Pedophobia is the fear of children

Pteronophobia is the fear of being tickled by feathers.

Pyrophobia is the fear of fire.

PLANTS

Botanophobia is the fear of plants

The plant life in the oceans makes up about 85 percent of all the greenery on the Earth.

PSYCHOLOGY

Agoraphobia is the fear of open spaces or crowds.

Aphenphosmphobia is the fear of being touched.

Atychiphobia is the fear of failure.

Autophobia is the fear of being alone.

Catagelophobia is the fear of being ridiculed.

Claustrophobia is the fear of confined spaces.

Scoptophobia is the fear of being stared at.

Xenophobia is the fear of strangers or foreigners.

RELIGION

According to Genesis 1:20-22, the chicken came before the egg.

SCARY

Cacophobia is the fear of ugliness

Hemophobia is the fear of blood.

Necrophobia is the fear of death or dead things

Nyctophobia is the fear of the dark.

Samhainophobia is the fear of Halloween

SCIENCE

The name of the radioactive element in smoke detectors in americium

Natural gas has no odor. They smell is added artificially so that leaks can be detected.

J is the only letter that doesn't appear anywhere on the periodic table of elements.

A 'jiffy' is an actual unit of time for 1/100 of a second.

There is no mercury on Mercury ... most of the planet is solid iron.

Most-used metals in the world (in order): iron, aluminum, copper

In 2012, Healthpoint Biotherapeutics, of Fort Worth, Texas developed a portable spray-on skin is actually made up of skin cells and suspended in a mixture of different types of proteins.

Vestergaard Frandsen is a company in Europe that launched a LifeStraw in 2006, which makes contaminated water safe to drink. Users simply just have to suck up the contaminated water through it as if they were any other regular straw. Without using batteries or chemicals. It removes 99.9% of parasites, 99.9999% of viruses and provides one person with an entire year's worth of drinking water for only $20.00.

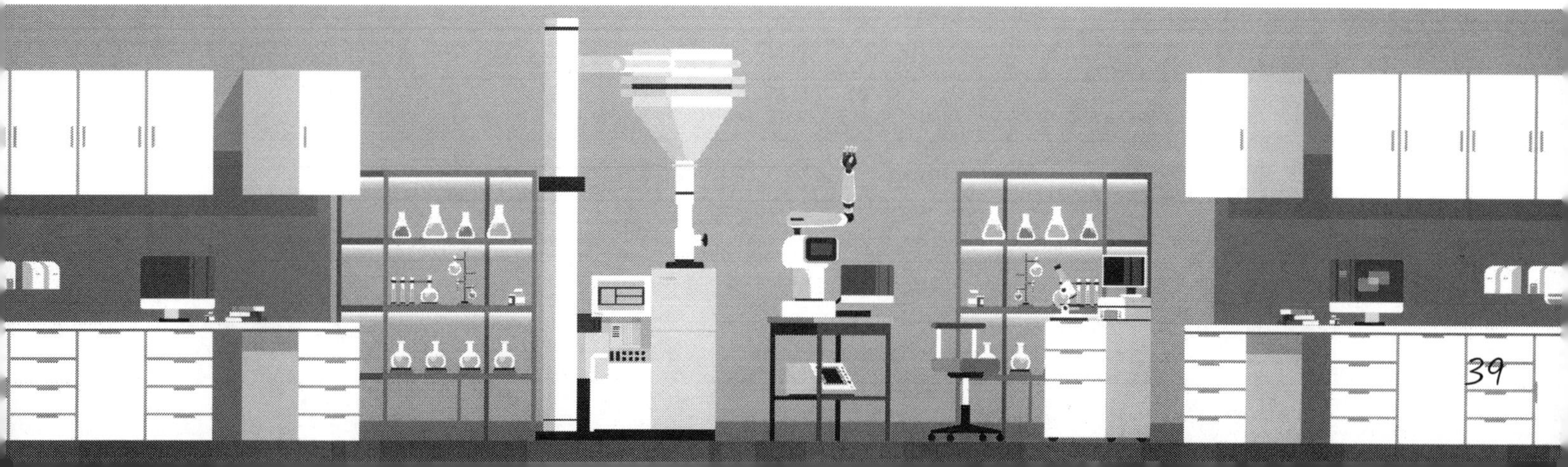

SLEEP

See Aircraft: Pilots sleeping.

Somniphobia is the fear of sleep.

See Animals: Snails Longest Nap

According to Daniel J. Buysse, a psychiatrist at the University of Pittsburgh Medical Center, about 2% of the population are considered the sleepless elite. This means that they are night owls and early birds simultaneously.

SOCIAL MEDIA

The graffiti artist David Choe, who painted Facebook's first office in 2005, was paid in shares and is now worth over $200,000,000.

Pope Francis broke a record when he hit one million Instagram followers within 12 hours of starting the account. The previous record was set by David Beckham at one million within 24 hours. He christened the account on a Saturday with a photo of him kneeling in prayer with an accompanying message that said "Pray for me," in nine languages.

More people of died from selfies than shark attacks.

SPACE

Jupiter is **bigger** that all the other planets in our solar system combined.

The **largest known star** (in terms of mass and brightness) is called the Pistol Star. It is believed to be 100 times as massive as our Sun, and 10,000,000 times as bright! In 1990, a star named the Pistol Star was known to lie at the center of the Pistol Nebula in the Milky Way Galaxy.

The Sun accounts for 99.86% of the mass in the solar system.

Uranus and Venus are the only planets that rotate clockwise. The other six planets in the solar system rotate counterclockwise.

SPORTS

Greatest Baseball Player of All Time is Babe Ruth who singlehandedly changed the sport of baseball. He became the first true superstar, having one of the most recognizable faces in American history. His memory lives on as if it is a folktale, being the first athlete who was seen as an American hero. Much of this was because he had the stats to back it up.

Most Home Runs All Time goes to Baseball Slugger Barry Bonds with 762.

Best Basketball Player of All Time. Kobe Bryant has more career points than Michael Jordan, but Jordan was the better scorer between the two. Highlighted by his NBA record 10 scoring titles, Jordan also averaged 30.1 points per game. ... He led the league in scoring 10 out of his 15 seasons and scored a career-high 37.1 PPG in 1986.

Best Cricket Player of All Time is Sir Donald Bradman, also known as The Don, is considered by most experts as the greatest batsman ever. **His batting average in test cricket was a staggering 99.94, an achievement that is unmatched in any major sport. Bradmanesque** is a term **synonymous with exceptional excellence**, not just in the world of cricket but also in other fields. He is considered as the epitome of excellence in cricket.

Greatest Football Player of All Time is Tom Brady. No player in the 101-year history of the National Football League has as many championships as Brady, who holds the record with seven, and five Super Bowl MVPs -- the most in NFL history -- in addition to owning numerous league records, cementing himself as the greatest of all time.

The Olympic flag's colors are always red, black, blue, green, and yellow rings on a field of white. This is because at least one of those colors appears on the falg of every nation on the planet.

Most **popular game** in the world is Soccer.

Best golf player of all time in the world is Jack Nicklaus. Nick-named the Golden Bear, Nicklaus gained 73 victories in his career, including 18 major golf championships. Nicklaus is the first player to win up to $2m in prize money, and, at 46 years old, he is the old-est player to win the Masters.

Most Decorated Olympian American swimmer, Michael Phelps is the youngest, having won 28 medals in both individual and team events (23 gold, 3 silver, and 2 bronze).

Highest Paid Athlete Roger Federer – USD$106.3 Million. Honorable mention to Cristiano Ronaldo – USD$105 Million, and Lionel Messi – USD$104 Million.

Best Rugby player of all time is Richie McCaw Where do we start, two-time World Cup-winning captain in 2011 and 2015, and played in the 2003, 2007. Named world rugby player of the year three times, in 2006, 2009 and 2010. A breakdown master, super athlete, inspirational leader, a remarkable ability to talk referees onto his side, and one of the most durable All Black player ever, McCaw had everything.

Best Soccer player of all time is Retired Brazilian football player Pele still considered one of the greatest soccer players ever.

The Guinness World Record for the **most goals** scored in a specified period is **1,281** by the Brazilian great Pele, born Edson Arantes do Nascimento. Pele scored these goals between September 7th, 1956,

SURVIVAL

To escape the grip of a crocodile's jaws, push your thumbs into its eyeballs – it will let you go instantly.

TREES

Leonardo da Vinci was the first to record that the number of rings in the cross-section of a tree trunk reveals its age. He also discovered that the width between the rings indicates the annual moisture.

VOCABULARY

Hippopotomonstrosesquipedaliophobia is the fear of long words

The only three English words that end in 'ceed': succeed, proceed and exceed.

"Dreamt" is the only English word that ends in the letters "mt".

The words 'flammable' and 'inflammable' mean the same thing

Longest word *pneumonoultramicroscopicsilicovolcanoconiosis*

Appearing in the Oxford English Dictionary, this 45-letter word for a disease is the longest English word that is defined in a major dictionary. It's a technical word referring to the lung disease more commonly known as silicosis

The **longest one-syllable** word in the English language: 'screeched'.

The words 'loosen' and 'unloosen' mean the same thing

The word "news" came from the first letters of the words North, East, West, and South. This was because information was being gathered from all different directions.

Only word in the English language that begins and end with 'und': underground.

There are only three words in English with the letters 'uu' are vacuum, residuum, and continuum.

Pandiculating is when you yawn and stretch at the same time.

The longest one-syllable word in the English language is "screeched".

Stewardess is the longest word in the English language typed with just the left hand.

On average, a woman utters around 7,000 words a day while a man uses iust over 2.000.

WEATHER

-40 Celsius and -40 Fahrenheit are the same temperature.

Astraphobia is the fear of thunder and lightning

Lilapsophobia is the fear of hurricanes and tornadoes.

You can determine the temperature outside by counting the chirps made by crickets. According to the Farmer's Almanac, to convert cricket chirps to degrees Celsius, you count the number of chirps in 25 seconds, then divide by three, then add four to get the temperature.

WELLNESS

Pathophobia is the fear of disease

Dentists have recommended that a toothbrush be kept at least 6 feet away from a toilet to avoid airborne particles resulting from the flush.

Dentophobia is the fear of dentists

Mysophobia is the fear of dirt and germs,

Iatrophobia is the fear of doctors.

Nosocomephobia is the fear of hospitals

Hypochondria is the fear of illness

Algophobia is the fear of pain.

Exercising increases productivity. Having a regular exercise routine can make you happier, smarter and more energetic and thus last longer throughout the day.

Right-handed people live, on average, nine years longer than left-handed people.

Obesophobia is the fear of gaining weight.

About Kenny

Kenny 'K3' Rochon, III is a 7-year-old boy, Game Changer - Leader. He recently was published as an author, and he is an expert in children's jokes.

When K3 is not telling jokes, he is playing chess with his daddy, learning how to code, playing Minecraft and Nintendo, and even some learning games that challenge his inquisitive and problem-solving mind.

He enjoys math, playing Sudoku, and expanding his vocabulary with his "Kenny's Fortune" word box. Kenny's Future Word Box is a future product he is coming out with the help other children learn the military alphabet and learn positive powerful words. He uses these words to hone up on his skills as a proficient speller so he can win spelling bees.

He has a yellow belt in Tae Kwon Do and practices so he can protect his Mommy (and Daddy) from danger.

About Ken Rochon

Ken loves his son Kenny. He enjoys writing a new book every year to celebrate his son's life and to share valuable lessons with him on character, leadership, and success

You can listen to Ken on his radio show Amplified with Ken Rochon, through the VoiceAmerica Influencer Channel on Mondays at 11 AM Eastern, on Itunes or at www.AmplifiedRadioShow.com

This is Ken's 28th book and now his favorite :) He wants to inspire other parents to create a legacy of love with their children through publishing. Ken has created templates that make publishing books more effortless and affordable. (www.PerfectPublishing.com)

Besides playing with his son, Ken loves to travel the world and has been to over 100 countries. He is a world class photographer and enjoys covering the biggest events in the USA (www.BIGeventsUSA.com)

If you are interested in having Ken speak at your school or event, simply connect with him on Facebook or email him at Ken@TheUmbrellaSyndicate.com.

Ken is the co-founder of the Keep Smiling Movement.com. He has co-authored a series of books focused on setting your mindset to be positive. Check out on Amazon the Keep Smiling - Shift Happens Series.

He resides in the Baltimore Washington Metropolitan Area with his family.

More of Kenny's Books

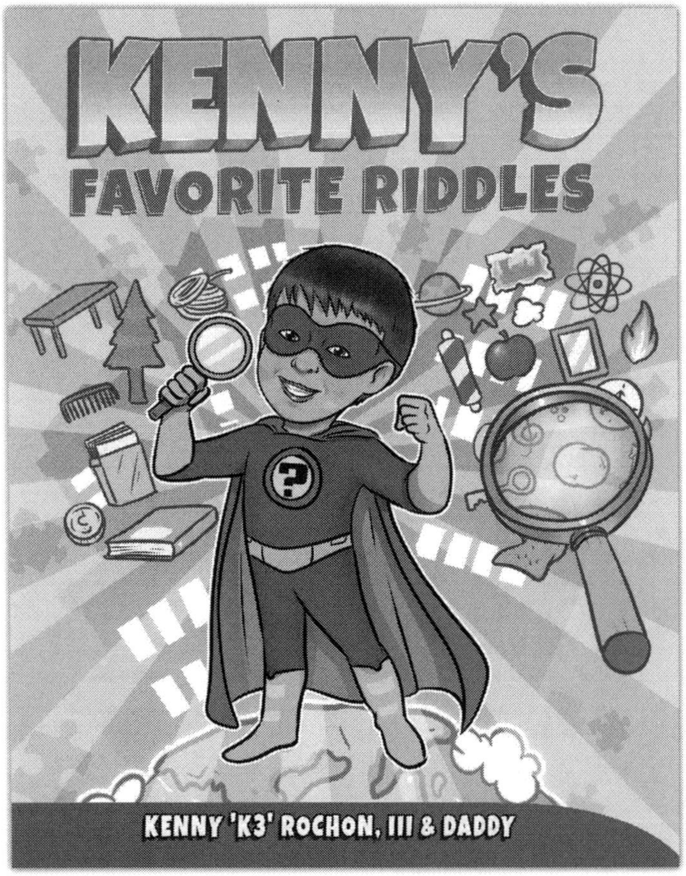

More of Kenny's Books

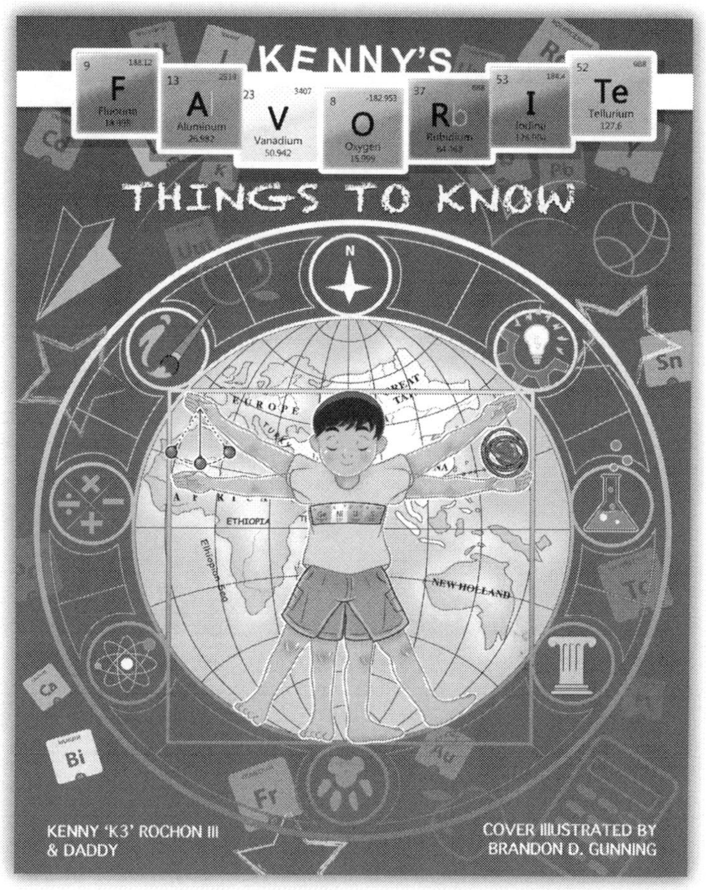

Made in United States
Orlando, FL
04 November 2026

90203686R00033